Math Comprehension Workbook

ESSENTIAL BASICS

PAUL GARNER

Published by:

PANTHERA PUBLISHING
PANTHERAPUBLISHING.COM
STUART, FLORIDA

Table of Contents

INTRODUCTION

How to Use This Workbook

Welcome to this **Math Workbook**! We're excited to embark on this journey with you and your child. Whether you're a parent helping at home or a student eager to strengthen your math skills, this workbook is designed to provide an effective, engaging, and systematic approach to mastering the core operations of addition, subtraction, multiplication, and division. These foundational skills are essential for succeeding in math and navigating real-world situations where numbers play an important role.

Why This Workbook is Important

Math is more than just numbers and equations—it's a way of thinking and solving problems. At this stage in your child's education, building a strong foundation in basic arithmetic is critical for future success. Math concepts will only grow in complexity as your child progresses through school. A firm grasp of addition, subtraction, multiplication, and division allows students to approach more advanced topics with confidence and help them finish sooner.

The Importance of Practice

Bill Gates is often perceived as an overnight success due to his immense accomplishments with Microsoft. Gates had access to a computer terminal at a young age (a rare opportunity in the 1960s) and spent thousands of hours coding and experimenting. His regular practice equipped him to seize the moment when personal computers became a revolutionary industry.

The Beatles are often regarded as one of the greatest musical bands in history, but their success was built on years of intense practice and performance. They played for hours each day in clubs. By the time they burst onto the global music scene, The Beatles had already accumulated over 1,200 live performances. For context, most modern bands don't perform that many shows in their entire careers.

Michael Jordan famously didn't make his high school varsity basketball team in his sophomore year, but he used the setback as motivation to practice intensively. Over years of deliberate and focused practice, he became one of the greatest athletes in history. This shows that natural talent alone isn't enough—consistent, targeted effort is key.

Just as learning to play a musical instrument or mastering a sport requires practice, so does developing math proficiency. Frequent practice:

Builds Confidence: Regularly working on problems helps your child realize that they *can* solve them, boosting their self-esteem.

Develops Familiarity: As your child practices, they will recognize patterns and strategies, making problem-solving feel more natural.

Encourages Retention: Skills practiced consistently are less likely to be forgotten.

Prepares for Advanced Math: Without a strong foundation in arithmetic, topics like fractions, decimals, percentages, and algebra can feel overwhelming. This workbook provides the building blocks necessary for tackling these future challenges.

How This Workbook is Structured

This workbook is divided into the four main sections, focusing on the key operations of arithmetic. These are used in all math activity in school and life.

Tips for Using This Workbook

To maximize the effectiveness of this workbook, we've included some tips for parents and students:

For Parents:

1. **Set a Routine:** Encourage your child to work on math problems daily, even if it's just for 10–15 minutes. Consistency is key.

2. **Provide a Supportive Environment:** Ensure your child has a quiet, distraction-free space to focus.
3. **Be Patient and Encouraging:** If your child struggles with a concept, remind them that making mistakes is part of the learning process. Offer guidance but allow them to figure out solutions independently whenever possible. (Answer keys are provided for each section.)
4. **Review Progress Together:** Celebrate small victories, such as mastering a challenging problem or completing a section. Positive reinforcement goes a long way in building confidence.

For Students:

1. **Take Your Time:** Don't rush through problems. Understanding the process is more important than speed.
2. **Practice Daily:** The more you practice, the easier math will become. Treat this workbook as a fun challenge rather than a chore.
3. **Use Scratch Paper:** Work out problems on scratch paper when needed. Writing things out helps with understanding.
4. **Ask for Help When Needed:** If something doesn't make sense, ask a parent, teacher, or sibling for assistance. There's no shame in seeking help.

Staying Motivated

Math can be challenging, but it's also incredibly rewarding. Celebrate progress, no matter how small. Completing a section or solving a tough problem are achievements worth recognizing. Keep in mind that every effort your child makes today is paving the way for success in the future.

A Strong Foundation for the Future

By working through this workbook, your child is building skills that extend far beyond math. Arithmetic teaches discipline, logical thinking, and problem-solving—all critical for academic success and life. A solid grasp of numbers will make future math methods, such as fractions, decimals, percentages, and algebra, much easier to learn. This foundation will serve your child well in school, their future career, and daily life.

ADDITION

Understanding the terms

In addition, the terms involved are called:

1. Addends: The numbers being added together.
    - Example: In 4+5 both 4 and 5 are addends.
2. Sum: The result of the addition.
    - Example: In 4+5 = 9, 9 is the sum.

Quick Example

For 7+3=10

- 7 is an addend.
- 3 is an addend.
- 10 is the sum.

This makes it easy to identify the parts of any addition problem!

How to Add

The first part of this chapter provides practice adding numbers that are stacked one on top of another. The second part will provide practice adding numbers that are shown on the same line.

For the stacked form it is very important to make sure that the numbers line up in each column. You will see that pattern in the problems below.

PROCESS

Step 1: Write the Numbers

- Write the numbers you want to add in a **stacked format**, one below the other.
- Make sure the digits are lined up correctly:
    - **Ones under ones**
    - **Tens under tens**
    - **Hundreds under hundreds**, and so on.

For example:

123

+456

The problems below have already been written out for you.

Step 2: Start with the Rightmost Column (Ones Place)

- Begin adding the numbers in the **ones place** (the far-right column).
- If the sum is **less than 10**, write the total below the line.
- If the sum is **10 or more**, write the **ones digit** below the line and carry the **tens digit** to the next column.

Example:

```
  123
 +456
 ____
    9    (3 + 6 = 9)
```

Step 3: Move to the Next Column (Tens Place)

- Add the digits in the **tens place**.

- Don't forget to **include the number you carried over** (if any) from the previous step.

Example:

```
  123
 +456
 ----
   79    (2 + 5 = 7)
```

Step 4: Move to the Next Column (Hundreds Place)

- Add the digits in the **hundreds place**, including any number you carried over.

Example:

```
  123
 +456
 ----
  579    (1 + 4 = 5)
```

Step 5: Write the Final Answer

- Once all the columns are added, the number below the line is your **final answer**.

Key Tips

1. Always line up your numbers by their place values.
2. Start adding from the **right (ones place)**.
3. Carry over numbers when the sum is 10 or more.
4. Double-check your work to make sure everything adds up correctly.

This process works for any numbers, big or small! 🎉

Practice

1)	2)	3)	4)	5)
52	741	824	474	830
+ 464	+ 757	+ 217	+ 869	+ 416
______	______	______	______	______

6)	7)	8)	9)	10)
467	256	548	630	713
+ 355	+ 808	+ 542	+ 97	+ 265
______	______	______	______	______

11)	12)	13)	14)	15)
66	960	651	376	222
+ 282	+ 333	+ 369	+ 333	+ 220
______	______	______	______	______

ADDITION

16)	17)	18)	19)	20)
428	249	594	87	161
+ 621	+ 689	+ 606	+ 54	+ 253
______	______	______	______	______
21)	22)	23)	24)	25)
404	294	899	393	258
+ 588	+ 838	+ 531	+ 12	+ 171
______	______	______	______	______
26)	27)	28)	29)	30)
648	318	879	574	614
+ 981	+ 248	+ 138	+ 610	+ 696
______	______	______	______	______

31)	32)	33)	34)	35)
851	400	275	402	438
+ 106	+ 824	+ 681	+ 836	+ 837
____	____	____	____	____

36)	37)	38)	39)	40)
226	900	982	872	654
+ 733	+ 225	+ 966	+ 527	+ 462
____	____	____	____	____

41)	42)	43)	44)	45)
164	616	373	510	381
+ 646	+ 855	+ 62	+ 64	+ 456
____	____	____	____	____

46)	47)	48)	49)	50)
305	22	617	39	320
+ 968	+ 653	+ 537	+ 719	+ 981
______	______	______	______	______

Answer Key

1) 516	2) 1498	3) 1041	4) 1343	5) 1246
6) 822	7) 1064	8) 1090	9) 727	10) 978
11) 348	12) 1293	13) 1020	14) 709	15) 442
16) 1049	17) 938	18) 1200	19) 141	20) 414
21) 992	22) 1132	23) 1430	24) 405	25) 429
26) 1629	27) 566	28) 1017	29) 1184	30) 1310
31) 957	32) 1224	33) 956	34) 1238	35) 1275
36) 959	37) 1125	38) 1948	39) 1399	40) 1116
41) 810	42) 1471	43) 435	44) 574	45) 837
46) 1273	47) 675	48) 1154	49) 758	50) 1301

Math problems are often laid out in a linear (line) layout. In cases like this, it is helpful to rewrite the problem in a stacked layout where you can visually see that the numbers are properly aligned as ones, tens, hundreds, etc.

For example,

A linear problem may look like this 616 + 945 =_______

This can be rewritten like this.

  616
+945

Using the same steps as the stacked practice exercises, the answer can be found.

Practice

1)	2)	3)
616 + 945 =_______	205 + 142 =_______	926 + 682 =_______
4)	5)	6)
224 + 655 =_______	897 + 453 =_______	133 + 542 =_______
7)	8)	9)
659 + 173 =_______	243 + 665 =_______	165 + 291 =_______
10)	11)	12)
263 + 722 =_______	880 + 687 =_______	614 + 254 =_______
13)	14)	15)
779 + 811 =_______	33 + 884 =_______	492 + 76 =_______
16)	17)	18)
351 + 897 =_______	780 + 463 =_______	629 + 936 =_______

19)	20)	21)
265 + 481 =_______	459 + 66 =_______	329 + 249 =_______
22)	23)	24)
530 + 985 =_______	368 + 900 =_______	479 + 632 =_______
25)	26)	27)
221 + 258 =_______	952 + 848 =_______	862 + 657 =_______
28)	29)	30)
953 + 468 =_______	400 + 162 =_______	404 + 539 =_______
31)	32)	33)
215 + 885 =_______	957 + 866 =_______	456 + 701 =_______
34)	35)	36)
135 + 771 =_______	744 + 73 =_______	569 + 102 =_______

37)	38)	39)
117 + 215 =_______	666 + 478 =_______	206 + 500 =_______
40)	41)	42)
381 + 859 =_______	733 + 93 =_______	304 + 278 =_______
43)	44)	45)
207 + 808 =_______	997 + 827 =_______	827 + 317 =_______
46)	47)	48)
592 + 992 =_______	762 + 460 =_______	564 + 121 =_______
49)	50)	51)
543 + 945 =_______	858 + 175 =_______	786 + 234 =_______
52)	53)	54)
214 + 792 =_______	168 + 628 =_______	696 + 537 =_______

Answer Key

1) 1561	2) 347	3) 1608	4) 879
5) 1350	6) 675	7) 832	8) 908
9) 456	10) 985	11) 1567	12) 868
13) 1590	14) 917	15) 568	16) 1248
17) 1243	18) 1565	19) 746	20) 525
21) 578	22) 1515	23) 1268	24) 1111
25) 479	26) 1800	27) 1519	28) 1421
29) 562	30) 943	31) 1100	32) 1823
33) 1157	34) 906	35) 817	36) 671
37) 332	38) 1144	39) 706	40) 1240
41) 826	42) 582	43) 1015	44) 1824
45) 1144	46) 1584	47) 1222	48) 685
49) 1488	50) 1033	51) 1020	52) 1006
53) 796	54) 1233		

SUBTRACTION

Understanding the terms

1. Minuend: The number you start with or the number being subtracted from.
    - Example: In 10−3, the 10 is the minuend.
2. Subtrahend: The number being subtracted.
    - Example: In 10−3 3 is the subtrahend.
3. Difference: The result of the subtraction.
    - Example: In 10−3=7 is the difference.

Quick Example

For 15−8=7 :

- 15 is the minuend.
- 8 is the subtrahend.
- 7 is the difference.

These terms help explain each part of a subtraction problem!

How to Subtract

Similar to the chapter on Addition, the first part of this chapter provides practice subtracting numbers that are stacked one on top of another. The second part will provide practice subtracting numbers that are shown on the same line.

For the stacked form it is very important to make sure that the numbers line up in each column. You will see that pattern in the problems below.

PROCESS

Step 1: Write the Numbers

- Write the numbers you want to subtract in a **stacked format**, one below the other.
- Make sure the digits are lined up correctly:
    - **Ones under ones**
    - **Tens under tens**
    - **Hundreds under hundreds**, and so on.

For example:

```
  456
- 123
```

The problems below have already been written out for you.

Step 2: Start with the Rightmost Column (Ones Place)

- Begin subtracting the numbers in the **ones place** (the far-right column).
- Write the answer below the line.
- Example:
- 456
- - 123
- 3 *(6 – 3 = 3)*

Step 3: Move to the Next Column (Tens Place)

- Subtract the digits in the **tens place**.
- Don't forget to **include the number you carried over** (if any) from the previous step.

Example:

$$\begin{array}{r} 456 \\ \underline{-\ 123} \\ 33 \end{array} \qquad \textit{(5 - 2 = 3)}$$

Step 4: Move to the Next Column (Hundreds Place)

- Add the digits in the **hundreds place**, including any number you carried over.

Example:

Example:

$$\begin{array}{r} 456 \\ \underline{-\ 123} \\ 333 \end{array} \qquad \textit{(4 - 1 = 3)}$$

Step 5: Write the Final Answer

- Once all the columns are added, the number below the line is your **final answer**.

Key Tips

5. Always line up your numbers by their place values.
6. Start subtracting from the **right (ones place)**.
7. Double-check your work to make sure everything subtracts correctly.

This process works for any numbers, big or small! 🎉

Practice

1)	2)	3)	4)	5)
984	775	389	766	701
- 664	- 572	- 284	- 245	- 613
____	____	____	____	____
6)	7)	8)	9)	10)
959	818	470	663	780
- 195	- 663	- 77	- 108	- 461
____	____	____	____	____
11)	12)	13)	14)	15)
535	864	321	617	640
- 18	- 615	- 287	- 371	- 549
____	____	____	____	____

16)	17)	18)	19)	20)
659	944	944	307	728
- 160	- 691	- 773	- 158	- 260
______	______	______	______	______

21)	22)	23)	24)	25)
952	904	927	794	934
- 952	- 851	- 197	- 122	- 732
______	______	______	______	______

26)	27)	28)	29)	30)
909	501	601	353	730
- 464	- 178	- 381	- 45	- 86
______	______	______	______	______

31)	32)	33)	34)	35)
245	836	977	800	685
- 76	- 210	- 776	- 485	- 284
____	____	____	____	____

36)	37)	38)	39)	40)
346	443	645	843	471
- 89	- 248	- 337	- 435	- 453
____	____	____	____	____

41)	42)	43)	44)	45)
674	997	554	614	499
- 626	- 47	- 63	- 177	- 453
____	____	____	____	____

46)	47)	48)	49)	50)
989	754	900	866	517
- 603	- 293	- 848	- 332	- 217
_____	_____	_____	_____	_____

Answer Key

1) 320	2) 203	3) 105	4) 521	5) 88
6) 764	7) 155	8) 393	9) 555	10) 319
11) 517	12) 249	13) 34	14) 246	15) 91
16) 499	17) 253	18) 171	19) 149	20) 468
21) 0	22) 53	23) 730	24) 672	25) 202
26) 445	27) 323	28) 220	29) 308	30) 644
31) 169	32) 626	33) 201	34) 315	35) 401
36) 257	37) 195	38) 308	39) 408	40) 18
41) 48	42) 950	43) 491	44) 437	45) 46
46) 386	47) 461	48) 52	49) 534	50) 300

Practice

1)	2)	3)	4)	5)
854	667	871	400	621
- 181	- 283	- 736	- 106	- 116
____	____	____	____	____

6)	7)	8)	9)	10)
676	636	583	934	218
- 316	- 96	- 553	- 842	- 156
____	____	____	____	____

11)	12)	13)	14)	15)
812	850	780	588	526
- 718	- 327	- 388	- 377	- 479
____	____	____	____	____

SUBTRACTION

16)	17)	18)	19)	20)
776	333	867	168	249
- 590	- 269	- 266	- 166	- 184
______	______	______	______	______

21)	22)	23)	24)	25)
962	400	675	695	945
- 528	- 304	- 239	- 335	- 58
______	______	______	______	______

26)	27)	28)	29)	30)
677	998	463	722	873
- 280	- 385	- 371	- 263	- 783
______	______	______	______	______

31)	32)	33)	34)	35)
469	252	995	714	323
- 134	- 206	- 25	- 334	- 293
_____	_____	_____	_____	_____

36)	37)	38)	39)	40)
329	925	765	633	780
- 157	- 482	- 48	- 83	- 690
_____	_____	_____	_____	_____

41)	42)	43)	44)	45)
762	937	330	653	879
- 411	- 181	- 78	- 306	- 169
_____	_____	_____	_____	_____

SUBTRACTION

46)	47)	48)	49)	50)
440	927	989	768	213
- 171	- 472	- 437	- 716	- 173
______	______	______	______	______

Answer Key

1) 673	2) 384	3) 135	4) 294	5) 505
6) 360	7) 540	8) 30	9) 92	10) 62
11) 94	12) 523	13) 392	14) 211	15) 47
16) 186	17) 64	18) 601	19) 2	20) 65
21) 434	22) 96	23) 436	24) 360	25) 887
26) 397	27) 613	28) 92	29) 459	30) 90
31) 335	32) 46	33) 970	34) 380	35) 30
36) 172	37) 443	38) 717	39) 550	40) 90
41) 351	42) 756	43) 252	44) 347	45) 710
46) 269	47) 455	48) 552	49) 52	50) 40

Math problems are often laid out in a linear (line) layout. In cases like this, it is helpful to rewrite the problem in a stacked layout where you can visually see that the numbers are properly aligned as ones, tens, hundreds, etc.

For example,

A linear problem may look like this 951 - 760 =_______

This can be rewritten like this.

$$\begin{array}{r} 951 \\ \underline{-\ 760} \end{array}$$

Using the same steps as the stacked practice exercises, the answer can be found.

1)	2)	3)
951 - 760 =_______	687 - 351 =_______	793 - 45 =_______
4)	5)	6)
796 - 488 =_______	606 - 214 =_______	945 - 909 =_______
7)	8)	9)
656 - 24 =_______	965 - 665 =_______	954 - 19 =_______
10)	11)	12)
300 - 82 =_______	809 - 586 =_______	64 - 49 =_______
13)	14)	15)
568 - 321 =_______	514 - 418 =_______	736 - 456 =_______
16)	17)	18)
329 - 83 =_______	577 - 414 =_______	765 - 584 =_______

19)	20)	21)
809 - 655 =_______	799 - 422 =_______	723 - 696 =_______
22)	23)	24)
733 - 200 =_______	775 - 79 =_______	950 - 255 =_______
25)	26)	27)
821 - 227 =_______	244 - 192 =_______	926 - 264 =_______
28)	29)	30)
717 - 263 =_______	775 - 668 =_______	761 - 315 =_______
31)	32)	33)
826 - 721 =_______	897 - 230 =_______	557 - 362 =_______
34)	35)	36)
362 - 212 =_______	633 - 583 =_______	148 - 70 =_______

37)	38)	39)
840 - 637 =_______	644 - 430 =_______	500 - 195 =_______
40)	41)	42)
973 - 771 =_______	489 - 386 =_______	677 - 551 =_______
43)	44)	45)
651 - 47 =_______	900 - 346 =_______	431 - 126 =_______
46)	47)	48)
773 - 364 =_______	750 - 417 =_______	722 - 442 =_______
49)	50)	51)
511 - 181 =_______	285 - 96 =_______	979 - 290 =_______
52)	53)	54)
621 - 313 =_______	853 - 325 =_______	797 - 182 =_______

Answer Key

1) 191	2) 336	3) 748	4) 308
5) 392	6) 36	7) 632	8) 300
9) 935	10) 218	11) 223	12) 15
13) 247	14) 96	15) 280	16) 246
17) 163	18) 181	19) 154	20) 377
21) 27	22) 533	23) 696	24) 695
25) 594	26) 52	27) 662	28) 454
29) 107	30) 446	31) 105	32) 667
33) 195	34) 150	35) 50	36) 78
37) 203	38) 214	39) 305	40) 202
41) 103	42) 126	43) 604	44) 554
45) 305	46) 409	47) 333	48) 280
49) 330	50) 189	51) 689	52) 308
53) 528	54) 615		

MULTIPLICATION

How to Do Multiplication: A Step-by-Step Guide

Multiplication is the process of adding a number to itself multiple times. It's like repeated addition. Here's how to multiply two numbers step by step:

Step 1: Understand the Terms

In multiplication, the terms involved are called:

1. Multiplicand: The number being multiplied.
    - Example: In 5×35 5 is the multiplicand.
2. Multiplier: The number you multiply by.
    - Example: In 5×3 3 is the multiplier.
3. Product: The result of the multiplication.
    - Example: In 5×3=15 is the product.

    Quick Example

    For 7×4=28

    - 7 is the multiplicand.
    - 4 is the multiplier.
    - 28 is the product.

    These terms help clarify the parts of a multiplication problem!

Step 1: Write the Numbers

- Write the larger number (if any) on top and the smaller number below it.
- Align the numbers by their **place values**.

- Draw a line under them to separate the problem from the answer.

  For example:

  $$\begin{array}{r} 243 \\ \times\ 5 \\ \hline \end{array}$$

Step 2: Start Multiplying from the Rightmost Digit

- Start with the rightmost digit of the bottom number (the **ones place**).
- Multiply it by each digit of the top number, one at a time, starting from the right.
- Write each result **below the line.**

Example:

  - Multiply **5 × 3** = **15**. Write the **5** in the ones place and **carry over the 1** to the tens place.
  - Multiply **5 × 4** = **20**, then add the 1 you carried over. Write **21** (write the 1 and carry the 2).
- Multiply **5 × 2** = **10**, then add the 2 you carried over. Write **12**.

  So:

  $$\begin{array}{r} 243 \\ \times\ 5 \\ \hline 1215 \end{array}$$

Step 3: Handle Larger Numbers with Multiple Digits

If the bottom number has more than one digit, multiply each digit **separately**, then add the results.

Example: Multiply **243 × 12**:

1. Multiply the **2** (ones place):

```
  243
×  12
  486
```

2. Multiply the **1** (tens place). Add a **zero** to the end of the result because it's in the tens place:

```
  243
× 12
  486
 2430
```

3. Add the two results together:

```
  243
× 12
  486
+2430
 2916
```

Step 4: Write the Final Answer

After adding all the products, the result is your final answer.

Tips for Multiplication

1. **Align by Place Value**: Keep your numbers aligned correctly in rows and columns.
2. **Carry Numbers Carefully**: Always remember to carry over extra digits.
3. **Use Zeros**: When multiplying by a digit in the tens, hundreds, etc., always add the right number of zeros to the result.

4. **Double-Check Your Work**: Multiplication is tricky; check your work to avoid mistakes.

By following these steps, you can multiply any two numbers, no matter how large or small! 🎉

Practice

1)	2)	3)	4)	5)
84	90	42	42	62
× 74	× 30	× 59	× 28	× 66
____	____	____	____	____

6)	7)	8)	9)	10)
22	67	67	26	97
× 15	× 85	× 48	× 54	× 79
____	____	____	____	____

11)	12)	13)	14)	15)
44	59	13	90	20
× 53	× 70	× 68	× 40	× 48
____	____	____	____	____

16)	17)	18)	19)	20)
79	47	49	55	35
× 42	× 77	× 74	× 25	× 17
____	____	____	____	____

21)	22)	23)	24)	25)
98	65	86	81	37
× 35	× 37	× 38	× 25	× 16
____	____	____	____	____

26)	27)	28)	29)	30)
54	63	38	39	66
× 50	× 12	× 22	× 78	× 96
____	____	____	____	____

31)	32)	33)	34)	35)
84	64	91	64	79
× 59	× 27	× 38	× 63	× 74
____	____	____	____	____
36)	37)	38)	39)	40)
71	33	55	35	41
× 78	× 59	× 60	× 47	× 97
____	____	____	____	____
41)	42)	43)	44)	45)
56	51	43	68	27
× 62	× 12	× 22	× 23	× 43
____	____	____	____	____

46)	47)	48)	49)	50)
18	73	47	73	13
× 65	× 49	× 40	× 29	× 89
____	____	____	____	____

Answer Key

1) 6216	2) 2700	3) 2478	4) 1176	5) 4092
6) 330	7) 5695	8) 3216	9) 1404	10) 7663
11) 2332	12) 4130	13) 884	14) 3600	15) 960
16) 3318	17) 3619	18) 3626	19) 1375	20) 595
21) 3430	22) 2405	23) 3268	24) 2025	25) 592
26) 2700	27) 756	28) 836	29) 3042	30) 6336
31) 4956	32) 1728	33) 3458	34) 4032	35) 5846
36) 5538	37) 1947	38) 3300	39) 1645	40) 3977
41) 3472	42) 612	43) 946	44) 1564	45) 1161
46) 1170	47) 3577	48) 1880	49) 2117	50) 1157

Practice

1)	2)	3)	4)	5)
81	18	47	50	54
× 87	× 58	× 16	× 72	× 46
____	____	____	____	____

6)	7)	8)	9)	10)
95	63	65	26	11
× 98	× 57	× 18	× 35	× 23
____	____	____	____	____

11)	12)	13)	14)	15)
47	60	13	15	64
× 59	× 80	× 46	× 38	× 56
____	____	____	____	____

16)	17)	18)	19)	20)
63	91	61	75	15
× 39	× 79	× 48	× 24	× 31
____	____	____	____	____

21)	22)	23)	24)	25)
78	41	81	81	28
× 95	× 71	× 13	× 53	× 64
____	____	____	____	____

26)	27)	28)	29)	30)
91	68	69	10	16
× 95	× 11	× 96	× 95	× 36
____	____	____	____	____

31)	32)	33)	34)	35)
62	13	60	47	76
× 82	× 14	× 38	× 55	× 50
____	____	____	____	____

36)	37)	38)	39)	40)
39	46	78	68	47
× 16	× 61	× 62	× 82	× 81
____	____	____	____	____

41)	42)	43)	44)	45)
74	75	79	32	83
× 72	× 27	× 16	× 95	× 29
____	____	____	____	____

46)	47)	48)	49)	50)
26	93	33	69	78
× 31	× 33	× 74	× 79	× 37
______	______	______	______	______

Answer Key

1) 7047	2) 1044	3) 752	4) 3600	5) 2484
6) 9310	7) 3591	8) 1170	9) 910	10) 253
11) 2773	12) 4800	13) 598	14) 570	15) 3584
16) 2457	17) 7189	18) 2928	19) 1800	20) 465
21) 7410	22) 2911	23) 1053	24) 4293	25) 1792
26) 8645	27) 748	28) 6624	29) 950	30) 576
31) 5084	32) 182	33) 2280	34) 2585	35) 3800
36) 624	37) 2806	38) 4836	39) 5576	40) 3807
41) 5328	42) 2025	43) 1264	44) 3040	45) 2407
46) 806	47) 3069	48) 2442	49) 5451	50) 2886

Math problems are often laid out in a linear (line) layout. In cases like this, it is helpful to rewrite the problem in a stacked layout where you can visually see that the numbers are properly aligned as ones, tens, hundreds, etc.

For example,

A linear problem may look like 37 x 16 =_______

This can be rewritten like this.

$$\begin{array}{r} 37 \\ \underline{\times\ 16} \end{array}$$

Using the same steps as the stacked practice exercises, the answer can be found.

1)	2)	3)
37 x 16 =_______	54 x 50 =_______	63 x 12 =_______
4)	5)	6)
38 x 22 =_______	78 x 39 =_______	96 x 66 =_______
7)	8)	9)
84 x 59 =_______	64 x 27 =_______	91 x 38 =_______
10)	11)	12)
64 x 63 =_______	79 x 74 =_______	78 x 71 =_______
13)	14)	15)
59 x 33 =_______	60 x 55 =_______	47 x 35 =_______
16)	17)	18)
97 x 41 =_______	62 x 56 =_______	51 x 12 =_______

19)

43 x 22 =_______

20)

68 x 23 =_______

21)

43 x 27 =_______

22)

65 x 18 =_______

23)

73 x 49 =_______

24)

47 x 40 =_______

25)

73 x 29 =_______

26)

89 x 13 =_______

27)

51 x 13 =_______

28)

77 x 76 =_______

29)

84 x 28 =_______

30)

88 x 52 =_______

31)

84 x 46 =_______

32)

51 x 41 =_______

33)

82 x 32 =_______

34)

58 x 58 =_______

35)

66 x 17 =_______

36)

73 x 33 =_______

37)	38)	39)
34 x 15 =_______	96 x 39 =_______	68 x 42 =_______
40)	41)	42)
43 x 39 =_______	29 x 29 =_______	65 x 48 =_______
43)	44)	45)
71 x 31 =_______	64 x 63 =_______	17 x 14 =_______
46)	47)	48)
32 x 23 =_______	62 x 45 =_______	85 x 35 =_______
49)	50)	51)
90 x 57 =_______	44 x 10 =_______	32 x 24 =_______
52)	53)	54)
98 x 68 =_______	38 x 31 =_______	89 x 21 =_______

Answer Key

1) 592	2) 2700	3) 756	4) 836
5) 3042	6) 6336	7) 4956	8) 1728
9) 3458	10) 4032	11) 5846	12) 5538
13) 1947	14) 3300	15) 1645	16) 3977
17) 3472	18) 612	19) 946	20) 1564
21) 1161	22) 1170	23) 3577	24) 1880
25) 2117	26) 1157	27) 663	28) 5852
29) 2352	30) 4576	31) 3864	32) 2091
33) 2624	34) 3364	35) 1122	36) 2409
37) 510	38) 3744	39) 2856	40) 1677
41) 841	42) 3120	43) 2201	44) 4032
45) 238	46) 736	47) 2790	48) 2975
49) 5130	50) 440	51) 768	52) 6664
53) 1178	54) 1869		

DIVISION

How to Do Division: A Step-by-Step Guide

Division is the process of splitting a number into equal parts or groups. It helps us find out how many times one number fits into another. Here's how to divide step by step:

Step 1: Understand the Terms

- **Dividend**: The number you are dividing (the larger number).
- **Divisor**: The number you are dividing by (the smaller number).
- **Quotient**: The result of the division.
- **Remainder**: What's left over if the division is not exact.

  Example: In 25÷4

  - 25 is the **dividend**.
  - 4 is the **divisor**.
  - The answer will be the **quotient** with or without a remainder.

Step 2: Set Up the Long Division

1. Write the **dividend** inside a long division bracket.
2. Write the **divisor** outside the bracket.

   Example:

   $$4\overline{)\,25}$$

Step 3: Divide the First Digit

- Look at the first digit (or group of digits) of the dividend.
- Ask: How many times does the divisor fit into this digit **without exceeding it**?
- Write that number (the quotient) above the dividend.

Example:

4 fits into 25 **6 times** because 4×6=24

Write:

```
   _6___
4 | 25
```

Step 4: Multiply and Subtract

1. Multiply the divisor by the number you just wrote in the quotient.
2. Subtract this product from the current digits of the dividend.

   Example:

   25−24=1

   Write:

```
   _6___
4 | 25
   -24
     1
```

Step 5: Bring Down the Next Digit

If there are more digits in the dividend, bring down the next digit and repeat Steps 3–4.

If the dividend was larger, like 258:

1. After subtracting 24, bring down the next digit 8, making it 18.
2. Divide again: 4 fits into 18 **4 times**.

Step 6: Write the Final Answer

- The numbers written above the dividend form the **quotient**.
- If there's anything left that the divisor cannot divide, this is the **remainder**.

Example:

- $25 \div 4 = 6$ with a remainder of 1, or 6R1

Key Tips

1. **Estimate First**: If you're unsure how many times the divisor fits, estimate and adjust if needed.
2. **Check Your Work**: Multiply the divisor by the quotient and add the remainder to confirm it matches the dividend.
3. **Practice with Remainders**: Division often leaves remainders, so get comfortable with them.

With practice, division becomes easier to understand and solve!

Practice

1)	2)	3)	4)	5)
783	480	588	400	68
÷ 9	÷ 5	÷ 6	÷ 5	÷ 4
______	______	______	______	______
6)	7)	8)	9)	10)
558	656	744	168	102
÷ 9	÷ 8	÷ 8	÷ 2	÷ 3
______	______	______	______	______
11)	12)	13)	14)	15)
182	297	240	84	192
÷ 2	÷ 9	÷ 6	÷ 3	÷ 3
______	______	______	______	______

16)	17)	18)	19)	20)
522	56	357	340	388
÷ 6	÷ 2	÷ 7	÷ 4	÷ 4
____	____	____	____	____
21)	22)	23)	24)	25)
405	512	248	434	170
÷ 5	÷ 8	÷ 4	÷ 7	÷ 2
____	____	____	____	____
26)	27)	28)	29)	30)
46	196	435	180	42
÷ 2	÷ 4	÷ 5	÷ 2	÷ 3
____	____	____	____	____

31)	32)	33)	34)	35)
104	210	340	90	390
÷ 8	÷ 3	÷ 4	÷ 3	÷ 6
_____	_____	_____	_____	_____

36)	37)	38)	39)	40)
114	396	52	486	465
÷ 6	÷ 9	÷ 4	÷ 9	÷ 5
_____	_____	_____	_____	_____

41)	42)	43)	44)	45)
236	166	295	115	203
÷ 4	÷ 2	÷ 5	÷ 5	÷ 7
_____	_____	_____	_____	_____

46)	47)	48)	49)	50)
117	380	98	488	288
÷ 3	÷ 5	÷ 2	÷ 8	÷ 8
____	____	____	____	____

Answer Key

1) 87	2) 96	3) 98	4) 80	5) 17
6) 62	7) 82	8) 93	9) 84	10) 34
11) 91	12) 33	13) 40	14) 28	15) 64
16) 87	17) 28	18) 51	19) 85	20) 97
21) 81	22) 64	23) 62	24) 62	25) 85
26) 23	27) 49	28) 87	29) 90	30) 14
31) 13	32) 70	33) 85	34) 30	35) 65
36) 19	37) 44	38) 13	39) 54	40) 93
41) 59	42) 83	43) 59	44) 23	45) 29
46) 39	47) 76	48) 49	49) 61	50) 36

Practice:

1)	2)	3)	4)	5)
132	426	264	273	332
÷ 2	÷ 6	÷ 4	÷ 3	÷ 4
______	______	______	______	______

6)	7)	8)	9)	10)
162	154	288	280	138
÷ 3	÷ 2	÷ 8	÷ 5	÷ 2
______	______	______	______	______

11)	12)	13)	14)	15)
162	104	891	497	75
÷ 9	÷ 8	÷ 9	÷ 7	÷ 3
______	______	______	______	______

16)	17)	18)	19)	20)
648	189	102	78	279
÷ 9	÷ 9	÷ 3	÷ 2	÷ 9
____	____	____	____	____

21)	22)	23)	24)	25)
588	166	224	132	132
÷ 7	÷ 2	÷ 8	÷ 2	÷ 2
____	____	____	____	____

26)	27)	28)	29)	30)
470	108	192	72	192
÷ 5	÷ 3	÷ 6	÷ 4	÷ 2
____	____	____	____	____

31)	32)	33)	34)	35)
336	255	376	301	155
÷ 7	÷ 5	÷ 8	÷ 7	÷ 5
____	____	____	____	____

36)	37)	38)	39)	40)
48	672	188	648	425
÷ 3	÷ 8	÷ 4	÷ 9	÷ 5
____	____	____	____	____

41)	42)	43)	44)	45)
92	495	520	384	148
÷ 4	÷ 9	÷ 8	÷ 4	÷ 4
____	____	____	____	____

46)	47)	48)	49)	50)
237	180	585	702	232
÷ 3	÷ 9	÷ 9	÷ 9	÷ 4
_____	_____	_____	_____	_____

Answer Key

1) 66	2) 71	3) 66	4) 91	5) 83
6) 54	7) 77	8) 36	9) 56	10) 69
11) 18	12) 13	13) 99	14) 71	15) 25
16) 72	17) 21	18) 34	19) 39	20) 31
21) 84	22) 83	23) 28	24) 66	25) 66
26) 94	27) 36	28) 32	29) 18	30) 96
31) 48	32) 51	33) 47	34) 43	35) 31
36) 16	37) 84	38) 47	39) 72	40) 85
41) 23	42) 55	43) 65	44) 96	45) 37
46) 79	47) 20	48) 65	49) 78	50) 58

Math problems are often laid out in a linear (line) layout. In cases like this, it is helpful to rewrite division problems in the long form layout where you can visually see that the numbers are properly aligned.

For example,

A linear problem may look like 32 / 6 or 32 ÷ 6 =_______

This can be rewritten like this.

$$6 \overline{\smash{)}\,32}$$

Using the same steps as the stacked practice exercises, the answer can be found.

1)	2)	3)
32 ÷ 6 =_______	14 ÷ 4 =_______	52 ÷ 6 =_______
4)	5)	6)
43 ÷ 8 =_______	75 ÷ 8 =_______	28 ÷ 7 =_______
7)	8)	9)
48 ÷ 4 =_______	20 ÷ 6 =_______	84 ÷ 5 =_______
10)	11)	12)
79 ÷ 7 =_______	16 ÷ 5 =_______	55 ÷ 5 =_______
13)	14)	15)
41 ÷ 5 =_______	24 ÷ 3 =_______	43 ÷ 4 =_______
16)	17)	18)
24 ÷ 6 =_______	13 ÷ 4 =_______	85 ÷ 7 =_______

DIVISION

19)

47 ÷ 6 =_______

20)

10 ÷ 5 =_______

21)

97 ÷ 9 =_______

22)

80 ÷ 9 =_______

23)

96 ÷ 3 =_______

24)

71 ÷ 5 =_______

25)

26 ÷ 8 =_______

26)

97 ÷ 5 =_______

27)

89 ÷ 9 =_______

28)

23 ÷ 8 =_______

29)

12 ÷ 7 =_______

30)

91 ÷ 8 =_______

31)

73 ÷ 4 =_______

32)

17 ÷ 4 =_______

33)

83 ÷ 8 =_______

34)

19 ÷ 5 =_______

35)

16 ÷ 5 =_______

36)

51 ÷ 3 =_______

37)

27 ÷ 7 =_______

38)

64 ÷ 8 =_______

39)

38 ÷ 9 =_______

40)

10 ÷ 3 =_______

41)

76 ÷ 7 =_______

42)

51 ÷ 6 =_______

43)

87 ÷ 7 =_______

44)

73 ÷ 7 =_______

45)

44 ÷ 8 =_______

46)

30 ÷ 9 =_______

47)

81 ÷ 4 =_______

48)

39 ÷ 7 =_______

49)

99 ÷ 5 =_______

50)

54 ÷ 4 =_______

51)

40 ÷ 8 =_______

52)

91 ÷ 6 =_______

53)

80 ÷ 5 =_______

54)

41 ÷ 3 =_______

Answer Key

1) 5.33	2) 3.5	3) 8.67	4) 5.38
5) 9.38	6) 4	7) 12	8) 3.33
9) 16.8	10) 11.29	11) 3.2	12) 11
13) 8.2	14) 8	15) 10.75	16) 4
17) 3.25	18) 12.14	19) 7.83	20) 2
21) 10.78	22) 8.89	23) 32	24) 14.2
25) 3.25	26) 19.4	27) 9.89	28) 2.88
29) 1.71	30) 11.38	31) 18.25	32) 4.25
33) 10.38	34) 3.8	35) 3.2	36) 17
37) 3.86	38) 8	39) 4.22	40) 3.33
41) 10.86	42) 8.5	43) 12.43	44) 10.43
45) 5.5	46) 3.33	47) 20.25	48) 5.57
49) 19.8	50) 13.5	51) 5	52) 15.17
53) 16	54) 13.67		

OTHER BOOKS BY PAUL GARNER

Reading Comprehension for Busy Surfers 9thGrade

Stories and questions to help early teens learn to read for comprehension. Book contains short stories and longer stories followed by questions to help the teen learn to focus and question what they have read.

Reading Comprehension for Busy Dancers, 7th and 9th grades

7th and 9th graders have different books. Each has a collection of stories for busy dancers designed to help the student practice comprehension. Each story, tailored to their interests and grade, is followed by questions about the content of the stories to lead the student to think and comprehend what they read.

Reading Comprehension for Busy Athletes

A collection of stories for middle school and 9th grade athletes designed to help the student practice comprehension. Each story is followed by questions about the content of the stories to lead the student to think and comprehend what they read.

America Comprehension for Busy Patriots: Revolution

For the busy Patriot in middle school, this book combines reading comprehension with American history. Specifically, this book contains factual short stories of the revolutionary period in America's history from the Boston Massacre through the surrender of the British. It includes the heroic efforts of the "Sons of Liberty" and the leadership of George Washington. It also includes the influence of John Locke's ideas on the writing of the Declaration of Independence. These are exciting stories that also strengthen reading and vocabulary skills.

America Comprehension for Busy Patriots: Constitution

For the busy Patriot in middle school, this book combines reading comprehension with American history. Specifically, it contains factual short stories about the US Constitution from the Articles of Confederation through the Bill of Rights. It includes the philosophy about how a government of a free people should work, and the differences at the Constitutional Convention, the Federalist and Anti-Federalist papers, state ratification, as well as the first Constitutional crisis.

America Comprehension for Busy Patriots: Supreme Court

The United States Supreme Court stands as one of the most powerful and influential institutions in the country. Established in 1789 by the Constitution, the Supreme Court has the final say on matters of federal law and constitutional interpretation. This unique role makes the Court a critical guardian of justice, equality, and the rule of law in America.

The Supreme Court's primary responsibility is to interpret the Constitution and ensure that laws and government actions comply with constitutional principles. This power, known as judicial review, was established by the landmark case Marbury v. Madison in 1803. Judicial review allows the Court to invalidate laws and executive actions that it finds unconstitutional, thus serving as a check on the powers of the legislative and executive branches

This workbook contains the details of 17 of the most important and impactful cases decided by the court over 221 years. Each case is followed by comprehension questions leading students to consider carefully the importance of the case. Finally, there is an Answer Key in the back of the workbook.

Classical Comprehension for Busy Scholars: Greek Philosophers

The writings of classical philosophers significantly influenced the founders of America as they crafted the principles and structures of the new nation. Figures such as Thomas Jefferson, James Madison, and John Adams were deeply versed in classical philosophy and drew upon the ideas of thinkers like Plato and

Aristotle when developing the foundational concepts of democracy, governance, and individual rights. Aristotle's reflections on the rule of law and the importance of a balanced government inspired the American emphasis on checks and balances within the Constitution. Likewise, Plato's vision of a just society and the philosopher-king concept underscored the importance of wise and virtuous leadership, which resonated with the founders' aspirations for their new government. The emphasis on civic virtue and the common good found in the works of these philosophers provided a moral framework that guided the founders in their pursuit of liberty, equality, and justice. This philosophical grounding helped shape the American ethos, embedding values that continue to underpin the nation's democratic ideals and institutions.

Coming soon: Additional Greek, Roman, Christian and Enlightenment Philosophers

Yearning to Breathe Free: Seeking Liberty in a World of Authoritarians

Liberty for the people is resisted by those in power throughout the world. Most often liberty must be fought for, sometimes with violence, other times with persuasion, but never does it come naturally. Authoritarians can be found in families, churches, schools, businesses, governments, and communities always seeking to have their own way over the choices of others. Some are sociopaths with no conscience. Some are self-righteous moralists. Others may be driven by their own fears and their fears become a tyranny over others.

This book is about the quest for liberty in this world of authoritarians. It is not intended to be a book of history, but it does contain a great deal of history describing the many crossroads where liberty had chances to prevail but rarely did.

Made in the USA
Monee, IL
07 November 2026

67980868R00052